SPEAKING
TECHNICALLY

SPEAKING
TECHNICALLY

A Handbook for Scientists,
Engineers and Physicians on
How to Improve Technical Presentations

Sinclair Goodlad

Imperial College London, UK

Imperial College Press

Published by

Imperial College Press
57 Shelton Street
Covent Garden
London WC2H 9HE

Distributed by

World Scientific Publishing Co. Pte. Ltd.
5 Toh Tuck Link, Singapore 596224
USA office: 27 Warren Street, Suite 401-402, Hackensack, NJ 07601
UK office: 57 Shelton Street, Covent Garden, London WC2H 9HE

British Library Cataloguing-in-Publication Data
A catalogue record for this book is available from the British Library.

First published in 1990 by Sinclair Goodlad, Petersham Hollow
226 Petersham Road, Petersham, RICHMOND, Surrey TW10 7AL

Published in 1996 by Imperial College Press
Reprinted 2000, 2005

SPEAKING TECHNICALLY

ISBN 1-86094-034-X

Printed by FuIsland Offset Printing (S) Pte Ltd, Singapore

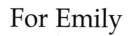

For Emily

Preface

If you

- have ideas and information to communicate;
- want to do so effectively;
- find the process stressful;
- have little time in which to prepare

then this booklet is for you.

The booklet is designed as a bag of tools to help you to tune up your presentation of complex information and ideas. To help in the run-up to a speaking assignment it offers

- headline points for rapid reference;
- brief commentary;
- review points and a check list for practice.

> **To use the book, read it straight through to get the line of argument. Then flip though the boxed-in headline points and the checklists when preparing your next presentation.**

A major part of this booklet (Chapter 3) is devoted to an analysis of strategies which (in my observation of technical presentations over some 25 years) can make speaking technically lively and memorable.

As a teacher, and provider of courses for university teachers and for scientists and engineers from industry and the civil service, I could see the need for a short, sharp, even provocatively directive booklet dealing, as this one does, with matters of immediate concern to professional people in a manner which, while not doing violence to the research literature, is not clogged up with references to it.

The booklet deals briefly, even dogmatically, with matters on which I draw upon my own observation of people speaking technically in after-dinner speeches; conference presentations; coping with visitors; inaugural lectures; industrial presentations; industrial tourism; lectures to students; presentations to funding bodies; project reports; public inquiries; school talks; section and departmental meetings; telephone calls and consultations.

The list of points in Appendix A is designed as a "flight check" to be used from the moment you receive an invitation to give a technical presentation. The notes in Appendix B, and the check list in Appendix C, are designed for use in Speaking Technically workshops which are indispensable if the thoughts about speaking technically offered here are to be translated into personal skills. I will be delighted to supply further information about workshops on request:

> **Sinclair Goodlad,**
> **226 Petersham Road, Petersham,**
> **Richmond, Surrey TW10 7AL.**
> **Telephone 020 8940 2800**
> **E-mail s.goodlad@imperial.ac.uk**

Finally, it is a great pleasure to thank some of the people without whose help this booklet would not exist: my wife, Inge, and my daughter, Emily, for their infinite patience; Joyce Brown, Michael Davis, Lewis Elton, Jane Gregory, Philip Healy and Barry Hill for commenting on the text; and Edward James for teaching me desk-top publishing. None of these bears any responsibility for the imperfections that remain.

Contents

Chapter 1

A Professional Job

Speaking technically is a key part of the work of any professional person. Indeed,

> **A professional person can be defined as someone who communicates to people less well-informed than the professional the opportunities for choice in technical matters.**

To do this, most professional people spend considerable amounts of time speaking technically — giving conference presentations, showing visitors around sites or laboratories, giving lectures to the lay public, making presentations to funding bodies, addressing juries or public inquiries, and so forth.

The art of speaking technically has nothing to do with acting, declamation, oratory, or the hype of salesmanship (which may have the object of subverting choice rather than encouraging it). Speaking technically is, however, one form of propaganda — the propagating of messages to influence action, for choices lead to

1

action. A basic axiom of this booklet is that

> **The object of speaking technically is always to change people's behaviour.**

Even when the only change in behaviour you intend is that your listeners should say 'Aha!' it is likely that you nevertheless want them to read a book or paper to follow up and build upon their initial excitement and interest. Therefore,

> **Always plan a talk with a view to what your listeners could do as a result of hearing it.**

You may have picked up this booklet because speaking technically makes you anxious. If you take the business seriously, you always will feel anxious — but anxious to do the best for your listeners, rather than anxious lest you make a fool of yourself. With proper preparation, you will not make a fool of yourself; and you are more likely to speak effectively if you follow the one key rule:

> **Think what your listeners need to know, not what you want to say.**

The thesis of the booklet is that in technical communication content matters most.

Get that right by thinking about the interests and needs of your listeners and your purpose in speaking technically, and you will feel less nervous. Any symptoms of nervousness that remain will not matter because your listeners will be too interested in what you are saying to notice.

It is, however, precisely because as a professional person you are sensitive to the needs of others, and perhaps because you may have suffered from people who were not, that you may find the whole prospect daunting. This is because

> **The most common fear in speaking technically is that of losing the audience's attention by boring those who know something about your subject or bewildering those who know little.**

The most difficult type of audience to address is that of mixed technical background. Indeed, because most audiences are of mixed background, Chapters 2 and 3 below deal with strategies for handling these matters.

If you are reasonably confident that you have planned the talk appropriately, your level of anxiety will be correspondingly reduced.

None of this is to imply that the presentation skills of voice production, stance, and so forth do not matter; they do (see Chapter 5 below). There can be little doubt that the authority of ideas is linked to the authority of the person who utters them. The key point in speaking technically, however, is that

> **You will feel authoritative if you know what you are trying to do and why; you will sound authoritative if you have thought about the process of communication as thoroughly as you have thought about the content.**

Most people have no difficulty explaining themselves in conversation, and can often make themselves both interesting and intelligible to fellow specialists and to non-specialists simultaneously. What is missing in formal presentations is immediate feed-back. You can get this by building question-answering time into your presentation (see Chapter 9 below); but you can substitute for the lack of direct feedback by very careful planning of your presentations.

There remains the anxiety of dealing with the unfamiliar.

> **Reduce your anxiety: Think what is the worst possible thing that could happen during your presentation and plan to prevent it.**

In short, try to turn a vague apprehension into a list of concrete and specific hazards, and then note down what action you would need to take. Here are some common ones:

Fear   : "I will forget what I was going to say."
Action : Have notes in an accessible format (see Chapters 5 and 6 below).

Fear   : "There will be someone there who knows more about the subject than I do."
Action : Be clear about the purpose of your presentation, and draw on others to help achieve it (see the section on Questions in Chapter 9 below).

Fear   : "I will be in a strange place, and things could go wrong."
Action : Check details as far ahead as possible, and have a "count-down" checklist for the day (see Appendix A).

With complex audio-visual equipment, use "belt and braces" by having back-up systems/materials (see Chapter 6 below).

Most of these points, like most of this booklet, will seem self-evident when spelt out. But when you are rightly and properly absorbed in the technical details of your specialism, you may overlook certain basic matters which could transform your presentation of ideas.

Chapter 2

Decide Your Purpose: Who Will Do What?

The most difficult, and most important, aspect of speaking technically is the initial leap of the imagination.

Your profession has achieved its status and respect by institutionalising certain ways of organising ideas and information. You personally have reached the point of being asked to give a technical presentation by paying careful attention to specific and minutely detailed observations and measurements. These observations and measurements gain their significance only from the intellectual framework in which they are perceived to be important.

The leap of imagination that is required in speaking technically is to judge the intellectual framework in which your listeners will be operating. This is why it is essential to

> **Think what your listeners need to know, not what you want to say.**

You need to climb back up the ladder from the particular details which absorb you in your professional role to the wider concepts and preoccupations which make the details interesting or important to a group wider than that of your immediate professional peers.

Audience interest and needs

Do not, however, jump off the top of the ladder! That is to say, do not strive for such a level of generality as to become meaningless or banal.

As your interest in certain details emerged from your preoccupations, so did the interests of your listeners also derive from their preoccupations. But these preoccupations may differ from yours. The key to effective technical communication is to identify precisely what the interests and preoccupations of your listeners are.

In speaking technically, there is no such thing as the "general public" or the talk for "general interest". If you find yourself using the word "general" you have probably failed to ask two key questions:

> **Who are my listeners? What are they going to do with the information I give them?**

In other words, how is their behaviour going to be changed?

Most forms of technical presentation are designed to encourage one of three basic behaviours:

- Find out
- Do
- Adjust.

All behaviours could be reduced to "DO" — but there are permutations and combinations that make even my threefold division pretty crude. However, the crucial point is that

> **Speaking technically must be for a specific audience for a specific purpose.**

Whether or not they will take immediate personal action, most people will have their interest caught if you assume that they will wish to find something out, do something, or adjust something they have already done. Chapter 3 examines strategies which point to action that people may take.

In most technical fields, these behaviours can be more readily represented as Research; Buy; Modify. There are others; but my object is to offer examples to help you plan your presentations rather than to be exhaustive.

To capture your listeners' interest, you must build around their existing knowledge and interests. Who are they? What are their experience and training, orientation, assumptions, attitudes, prejudices? What's in it all for them? Why should they want to hear a presentation on this subject? You need to find out.

You know from your own experience that the point of attachment to your subject was probably not purely intellectual; nor is it for most people. Our interest may come from sources that are not intrinsic to a given subject matter. To tap into this (sometimes subconscious) level of thought, ask yourself

> **What will people feel as a result of hearing my presentation?**

Aroused? More confident? Encouraged? Frightened? Inspired? More loyal to a group? Relieved?

It is most unusual for scientific or technological information to be presented "objectively" — as though the creators of the ideas lived in some sort of cultural amniotic sac. Most probably, there is a considerable investment not only of money but also of prestige, reputation, or organisational procedure in what you are doing. Try to tap into this.

Are there any emotions or predispositions which will make your job more difficult? How will you deal with them (suggestions in brackets):

- *Suspicion* (be level with people, put cards on the table);
- *Inertia* (get people excited);
- *Fear* (describe the source of fear to show that you understand it and sympathise);
- *Pride* (do not unnecessarily run down an existing system: indicate that you know why it was built that way and that you know too that people want to improve);
- *Incompetence* (treat people as if they have power to act: they may get power later even if they lack it at present).

If your audience is of mixed technical background, you will need to plan your presentation so as to interest and appeal to more than one level of attention. Don't patronise and don't bemuse. Assume that all your listeners are likely to be interested in those aspects of the subject that interest you — that they will want to go as far into the subject as they can. But at the same time, make sure you establish a framework of discourse that will bring those relatively new to the topic up to a level of comprehension that will keep them with you. Chapter 3 below offers detailed suggestions about how to do this. In short,

> **Never underestimate intelligence: never overestimate information.**

Always, it is essential to keep asking: what will people do as a result of hearing the presentation?

Lectures to students: a special case?

It is worth pausing here to reflect why lectures to students seem so deceptively easy to give and yet are so hard to give really well.

Students are tested before entry and might be thought to be relatively homogeneous in background, interests, and knowledge. In practice, even within a first-year class of highly-selected students, there is an astonishing range in all three. For this reason, if lectures are to be a significant component of a regime of instruction, it is particularly important to build in opportunities for the students to share ideas, for systematic feedback to the lecturer and for support activities so that the lecturing does not bear a load that it is not designed to carry.

What action will the students take? One needs to ask this question both of the short-term and of the long-term. It is easier to do the former. Immediately, the students may read a book or a paper; perform an experiment; discuss a concept in a seminar, etc.. In the longer-term, however, students will do many things: Research (research and development, design); Buy (procurement, installation, sales); Modify (management, civil service).

Herein lies the unique difficulty of teaching in higher education: the diffuseness of purpose.

It is possible to set some global objective for a degree course in terms of what graduates will do. For example, it is possible to set detailed instructional objectives on the assumption that a graduate scientist will be able to design, plan, and execute a complex experiment without detailed supervision, or that a graduate engineer will be skilled in project management. But, to avoid turning education into training (i.e. instruction for fixed purposes), one must teach as though a multiplicity of possible actions and careers might follow.

To ensure liveliness of presentation, I suggest that in series of lectures to students or other listeners you should

> **Plan each lecture, even when the lecture is part of a series or part of a complex system of instruction, to have its own inner logic.**

Chapter 3 offers suggestions on how to do this.

Pacing and timing

The courtesy of good timing in speaking technically is one of the qualities most highly valued by audiences; it is also quite difficult to achieve.

You have no doubt experienced the frustration of missing a train because someone was wittering on at a public meeting from which you could not escape, or the irritation of having your session shortened at a conference because the previous presenter went on too long, or the exasperation at having no chance to ask questions because all available time was taken up with talk. Bad timing often stems from one of two faults.

First, the speaker may simply not know how many words can reasonably be uttered per minute in different settings.

Second, the speaker may be tempted to "cover the ground", i.e. say everything — without proper discrimination about what matters most.

On the first point,

> **There is no definitive speed for speaking technically: speed must depend on content and setting. As a rough guide, 100 to 120 wpm is about right for a large audience and/or a presentation where special emphasis is needed; 150 wpm is probably the upper limit if you are to be understood.**

If you have little experience of speaking technically, you may wish to write out a complete text for your presentation to check the number of words you think you might need in the best of all possible worlds. But only do this if you then TEAR UP AND THROW AWAY YOUR SCRIPT.

Written prose has a much denser texture than spoken prose. If you work from a script, you will be hard to follow. Use the draft script to

> **Establish an outline of headings and subheadings and key facts and figures.**

But do not read it out (see Chapter 5 below on the use of notes). A script might help your timing; but it will murder your pacing!

Pacing is a function of structure; it has little to do with the number of words you speak per minute.

Our perception of pace depends on what we are looking and listening for. Think of the commentary on a horse-race: the commentator may be uttering words at a fantastic rate (maybe 350 wpm). But, the information content is low. We are listening only for the position of the wretched beast upon whose back our fortune rides.

By contrast, a technical presentation which does not have an adequate structure may be delivered very slowly and yet appear completely unintelligible.

The secret of pacing is not to overload your listeners or use technical presentations for purposes to which they are not suited.

> **Very frequently, the function of speaking technically is primarily that of spotlighting, highlighting, direction-finding, or map-making rather than that of conveying detailed information which your listeners must remember.**

If it is indeed your intention to use your technical presentation to highlight key ideas, you can sometimes afford to speak very quickly indeed. Just as time-lapse movie photography can point up broad patterns by speeding up a process (such as the growth of a plant or the movement of traffic through a road junction), so a technical presentation can help your listeners find their way through a mass of detail, the significance of which cannot readily be appreciated as details are encountered one by one.

For a high-speed, tour-of-the-highspots presentation, it is essential not only to tell your listeners your purpose and plan, but also to tell them what you expect them to do during and after your presentation. For example, you may be planning to go through material so quickly that it would be neither desirable nor possible for listeners to take notes. Your listeners will be able to relax, and thereby attend more intently to your presentation, if they are not worrying about what to do. Tell people what support materials you have provided or will be providing.

> **Presentations should not stand alone as the only method by which important information and ideas are communicated.**

Chapter 7 below examines the link between speaking technically and support materials and activities. Suffice it to note here that

> **You can speak very quickly, (thereby giving a sense of liveliness and excitement), without bewildering your listeners if you tell them what to look and listen for.**

Chapter 3

Planning The Content: The Frame of Reference

The audience for a technical presentation cannot go backwards and forwards in the way that the reader of a text can; yet each listener must know at every moment what is going on. It is, therefore, crucial to highlight the structure of your material when speaking technically.

> **Your listeners must know the principle of organisation of your presentation from the start — or they are likely to become lost and bewildered.**

A hoary old adage beloved of political speakers is: "Tell 'em what you're going to tell 'em; tell 'em; then tell 'em what you've told 'em." As a professional person interested in speaking technically, you may not want to be so crude. Yet, however complex your material, it is still worth trying to

> **Decide what is the single main point or message of your presentation.**

This is less difficult than it may at first seem — certainly if you accept that the aim of most speaking technically is to change people's behaviour. What is the most important experiment to be done (Research)? What unique virtue favours the system or device which you are seeking to commend (Buy)? What specific change in procedure would offer greatest benefit to your colleagues (Modify)?

The art of gaining attention is to cause your listeners to want actively the information and ideas you will be offering rather than to receive them passively.

Most procedures, techniques, systems, and devices in technology are designed to solve problems. In science, progress is often made through the solving of problems, the resolving of puzzles which validate or refute some system of thought. In many situations, a powerfully involving procedure is, therefore, to

> **Identify a problem, show how it might be tackled, go through the detail of the work, and finally review the significance of what has been done.**

If you are in doubt about what to emphasise, define your system, device, or process, thereby identifying its fundamental purpose. List the features which differentiate it from all other systems, devices, processes in the same broad category as yours. Decide which differentiating feature merits comment by asking what might be your listeners' purpose in wanting to know about it. And, as always, consider how their behaviour will change.

Example:

Telephone — a system or device for reproducing sounds at a distance which operates by converting the sounds into electrical impulses for transmission to a particular receiver.

This definition "locates" the entity within a very broad context intelligible to most school leavers. You can then pick on any one of a number of differentiating features for special emphasis. For example, "sounds". But what about pictures or data from computers? "Converting" in what way? Digital or analogue — and which is better for what purpose? "Transmission" — how? By wire? By radio? By optical fibre? "A particular receiver" — yes, but how to ensure that security is maintained?

The simple process of definition points up a multitude of issues which need to be elaborated if we are to appreciate the technical virtues of any particular system or device — say a cellular system for car telephones.

Once you have moved down the ladder from the broad categories to the specific categories, you can home in on some particular matter for detailed attention, if you

> **Use the process of defining a system to provide ideas with which to introduce it to non-specialists and make it simultaneously interesting to specialists.**

In the example given, you might decide that the unique value of your innovation was to produce a high-security system for sending facsimile documents to and from moving vehicles. Finance directors (or other listeners who are not specialists in engineering) would most likely be interested in the fact that such a system is technically possible — and also in its likely cost to the customer; by contrast, engineering colleagues might be interested in how, for example, you code or "scramble" input and output signals in a small, portable device to achieve security.

By defining your system or device, you can identify the hierarchy of ideas by which very detailed technical issues can be "located" within an appropriate framework. In taking an audience of mixed technical background through a presentation on this subject, you might make the focus or main point of your presentation the technique by which you had solved the security problem. In leading up to it, you might "remind" your listeners of the features which cellular telephones share with other telephonic systems, emphasising the advantages which the new system offers over previous systems.

Much of this might in fact be new to some of your listeners, but by "reminding" your audience as a whole, you would avoid talking down to the specialists. Having identified the key aspects of your new idea at the start, you put the specialists in a position to start thinking about how they would tackle the problem of security. Even if this does not keep them fruitfully occupied while you come down the conceptual ladder to the technicalities in which they are interested, they will at least have the satisfaction of appraising the way in which you build towards these technicalities.

> **The most common fault in speaking technically is to omit adequate introductory material. Be sure to 'locate' technical details within an appropriate frame of reference.**

If you do not state your purpose clearly, everyone will be left in uncertainty. Without the statement of plan, the specialists will get bored while you work your way towards the detail of your security system.

Clarity and organisation

Most of us can get whatever information we need if we work at it hard enough for long enough. The great value of hearing a professional person speaking technically is in the shedding of light where previously only darkness and confusion prevailed. In short,

> **What we value in hearing someone speaking technically is conceptual clarity — the presentation of principles of organisation and selection by which we can make informed choices.**

There is, however, great danger in simply presenting a system of classification as the primary focus of your presentation. It is a distressing paradox that the more powerful the system of classification is, the more boring it can be to hear it described in a lecture. What may make absorbing reading can be profoundly off-putting in a presentation. I have seen lecturers, whose clarity of thought is like a laser slicing through butter, lull their listeners into a trance by working their way systematically through some system of awesome power and lucidity.

The problem with such an approach is the very fertility of the content — there seems nothing for the listeners to do than absorb, sponge-like, the distilled drops of the speaker's wisdom. Their level of involvement is low precisely because the lucidity of presentation makes the ideas seem self-evident.

The proper place for systems of classification is in print. Presentations can then show people with case studies how they can be used. Therefore,

> **Never make a system of classification the focus of a presentation. Rather, use the system to address issues relevant to your listeners' interests and needs.**

The most important aspect of speaking technically is getting the order of ideas right by building around your listeners' interests.

Four types of order or structure commonly encountered in presentations are:

Time (and then, and then...narrative structure);

Space (left to right, top to bottom... topological structure);

Process (sequence of actions...may be a form of narrative organised by time, or by physical organisation of plant...functional structure);

Association (tenuous links between ideas...a chain of discourse).

All of these are primarily descriptive. There is nothing intrinsically wrong with them; indeed, order by association can become an art form as in Alastair Cooke's *"Letter from America"*. But

> **For speaking technically, descriptive forms of order are less involving than types of order which point to cause-effect relationships.**

Cause-effect relationships, in turn, are the more effective if they point to some action which may affect the lives (intellectual or material) of your listeners.

The selection of the appropriate principle of order or theme can give dramatic shape to your presentation.

The following pages offer a number of approaches of wide applicability to speaking technically. Each strategy is based on the supposition that some type of action by the listeners is the intended outcome:

Research : where the focus is on types of information needed for a specific purpose;

Buy : where the focus is on someone doing or getting something;

Modify : where the focus is on promoting change in the use or construction of some procedure, system, or device for greater economy, efficiency, reliability, safety, etc..

When you have decided what sort of presentation you are giving, i.e. what it is for and what you hope your listeners will do, you will be in a position to choose an appropriate theme to hold all the details together. For example, in a Research presentation, the detail of your work may be to validate a theory which is conceptually richer than another; your focus on the types of information needed might be to demonstrate the superior elegance and economy of your approach. That is your fundamental message or theme; the detail simply illustrates the thesis.

Similarly, with presentations designed to encourage people to Buy or Modify, the theme underlying the detail may be efficiency, safety, security, or reliability. Do not leave messages such as these to the end of a presentation, rather,

> **Help your listeners by stating your theme and purpose at the beginning of your presentation.**

For a mixed audience, i.e. containing some people with considerable technical knowledge and others with little, it is possible to develop two themes simultaneously. With the example of cellular telephones offered above, it would be possible to emphasise security as a theme for non-specialists and ingenuity as a theme for engineers. Many topics can be treated in this manner, but

> **If in doubt about which of two themes to develop, stick with the one best suited to those people in your audience who will take the most important action as a result of hearing your presentation.**

Insert enough additional material to retain the interest of others who may be present. The phrase "As you will know..."

will keep the experts content provided that you have given (in your introduction) promise of good things to come in terms of their interests.

Another example: both non-specialists and specialists may be interested in learning about High Definition Television (HDTV). Non-specialists may want to know roughly how it works, whether or not it really is better than the television system they already use, what it is likely to cost, and where it can be obtained. Specialists will also be interested in these matters, but their reason for listening to you may be to discover whether or not you have managed to solve a problem of camera-optics that is inhibiting progress. Both types of listener can benefit from your presentation provided that you

> **Explain at the outset the plan of your presentation as well as its purpose.**

They can then look and listen for those things that interest them. Buyers of the system might want to know if it can be manufactured at a cost competitive with existing systems; development engineers might be interested in the level of definition which can be achieved on a domestic receiver.

The technical specialists will not be bored by your scene-setting material if you identify clearly at the beginning the issue you propose to address that is at their technical level.

> **There is no such thing as a "general" audience.**

Every audience consists of individuals with specific, but often different reasons for wanting to hear what you have to say. You need to identify as clearly as possible what these reasons are and

> **Arouse the interest and attention of your listeners by giving them an incentive to attend to your presentation.**

There are many ways in which this necessary attention can be achieved. For purposes of illustration, the following pages examine types of dramatic structure which, if combined with proposals for action, can make presentations vivid:

1. Thesis

To build a presentation around a thesis challenges your listeners to evaluate and appraise information actively as you present it. Instead of wondering what is the significance of each item of information that you present, your listeners can judge whether it supports or weakens your thesis. Behind our enjoyment of this type of presentation is the satisfaction of watching someone trying to meet a challenge. Academic audiences (both faculty and students) thrive on disputatious modes of organisation — covertly thinking "Bet he/she can't prove that!" The thesis, being perhaps the most basic form of dramatic structure, can, however, be deployed in many contexts and circumstances. Indeed, the structures which follow can be considered to be but variations of it.

2. Disaster

Many people notice technology only when it goes wrong; indeed, the interest in science and technology shown by makers of radio and television programmes is often strongly linked with public perceptions of risk. You can, therefore, very fruitfully begin a technical presentation with a vivid evocation of a disaster. The body of the presentation can then examine, for example, design criteria which should have been followed to prevent the disaster; failure of designers to take account of one or more crucial factors in the failed system; benefit of the system you now propose which (of course!) embodies all the necessary criteria.

3. Problem

We don't normally scratch unless we have an itch; similarly, we are more likely to attend to a technical discourse if it addresses a question which we are anxious to have answered. The problem must, of course, be interesting and important to your listeners — and not artificially put together simply to illustrate a theoretical point. When you address an audience of mixed technical background, the art is to describe a problem that is interesting and meaningful to non-specialists and pick out for special treatment that aspect of it which will both satisfy the non-specialist and challenge the specialist.

4. Puzzle

A puzzle is very similar to a problem as a focus for a technical presentation. For practical purposes, I will regard a problem as a multi-dimensional, possibly even "fuzzy", question on which the organising concepts and ordering procedures of a professional discipline can throw light. By contrast, a puzzle is an unresolved anomaly or difficulty in the theory or procedure itself.

5. Comparison/Contrast

Differences are often best illustrated by comparisons or contrasts. Many technical presentations are designed to highlight the specific advantages of one system, device, idea, or procedure over others. A strategy based on comparisons or contrasts can be highly effective for this purpose.

6. Ninepins

A somewhat adventurous variation of comparisons and contrasts is the strategy in which you eliminate one after another of a set of competing items until only one (your preferred one) is left standing. The "ninepins" strategy can be particularly effective in a closed group (company, college, research group) in which all your listeners are ready to accept that the ideas or systems of competitors are nonsense — and are wanting you to highlight how and why they can be "rubbished".

These six dramatic structures for highlighting your themes can be combined with the three invitations to action to provide the following typology of strategies for many forms of speaking technically:

	Research	Buy	Modify
Thesis	A	B	C
Disaster	D	E	F
Problem	G	H	I
Puzzle	J	K	L
Comparison	M	N	O
Ninepins	P	Q	R

Strategy A: "I will argue that..."

Approach: *Research/Thesis*. It is no accident that the thesis is the form of intellectual organisation used for the highest level of academic qualification. The capacity to pull ideas and information together systematically to some purpose is highly valued; it is a more powerful indicator of intellectual competence than the capacity to merely accumulate information. Indeed, so deeply is the thesis mode of thinking embedded in western culture that it can be considered the primary form of technical argument.

Strategy A can be used whenever you wish to show that the gathering of certain information would further some object desired by your listeners.

Strategy B: "You should buy x because..."

Approach: *Buy/Thesis*. Strategy B can be deployed whenever you wish to demonstrate that certain information provides a conclusive argument for acquiring one system or device rather than another. It is a powerful strategy to use whenever some very complex thinking leads towards a certain course of action.

For example, you might wish to demonstrate that there are limits to the capacity of conventional computing systems to process information. Your thesis would be that parallel-processing is the way forward: the body of your presentation would explain why this is so. You might concentrate on explaining the physical limits of components in micro-circuits as the way of supporting your thesis.

Strategy C: "You should change procedure x because..."

Approach: *Modify/Thesis*. This strategy is very similar to B, the difference being that you might be demonstrating why some change to an existing system, device, or procedure would be beneficial.

Strategy C is appropriate to "in house" presentations where you are not necessarily trying to persuade people to buy something (because they have already bought it), but where you are trying to demonstrate that changes are needed in the way things are done. The thesis approach identifies clearly what modification is needed and then takes people through the reasoning behind your proposal.

Strategy D: "The evidence suggests that...but..."

Approach: *Research/Disaster*. The essential message in Strategy D is that "We do not yet know...". The method is to describe some disaster and to analyse the possible reasons for its having happened. To indicate the type of research needed, you might concentrate on showing what the existing evidence suggests, but go on to explain why further research is needed to get to the bottom of the matter.

Strategy E: "Don't let this happen to you..."

Approach: *Buy/Disaster*. This is a classic sales technique in which the strategy is to evoke images of doom, destruction, and decay until your listeners cry aloud: "What shall we do to be saved?" You then explain how they can avoid trouble by purchasing the equipment or services that you offer.

Strategy F: "It is urgent that we…"

Approach: *Modify/Disaster*. The strategy is similar to strategy E in that you evoke some painful scene to capture your listeners' interest and attention and then show the road to salvation. Unlike the approach in strategy D, (where you argue that further research is needed), here you show that we do know what went wrong and why, and to prevent future disasters — epidemics, system failures, misuse of devices, and so forth — you recommend specific action.

Strategy G: "We are nearly there…"

Approach: *Research/Problem*. This is a particularly valuable strategy to use to impress people! Rather than crowing about all the many virtues of the system, device, or procedure you want your listeners to adopt, you focus on the outstanding problem that your research is designed to crack. In building towards this, you mention in passing that you have long since solved all the major problems with which your competitors are still grappling.

You might, for example, explain how your research was designed to achieve the highest theoretically possible level of output or performance — say at the 85% level. In doing so you would demonstrate in a relaxed manner that you had left standing your competitors who were still struggling to achieve an extra five per cent efficiency at the 45% level.

Strategy H: "If we can solve this, then we should be able to…"

Approach: *Buy/Problem*. This strategy applies to all issues where the solution of a problem would have a very wide range of applications. It can be used where the object is to get funding for an area of applied research.

Strategy I: "An unexpected problem has arisen..."

Approach: *Problem/Modify*. This is a strategy to employ whenever you wish to recommend a change in a procedure. The procedure can be an administrative one and/or a technical one. The technique is to describe the way(s) in which the existing procedure is falling short of what is required, examine alternative approaches, and then recommend modifications which will achieve the desired object — of economy, reliability, security, or stability.

Strategy J: "Can you help us to solve this..."

Approach: *Research/Puzzle*. This is the classic strategy to use with an audience of potential recruits — to higher education institutions, industry, or research establishments. Rather than describing what has already been achieved, the technique is to describe what still needs to be found out. The art is to bring the listeners as near as possible to the cutting edge of the discipline by concentrating on a specific puzzle that lies at the heart of the discipline's concerns. The strategy is also a very effective basis for a lecture within a course in higher education.

Strategy K: "We need funds to crack this..."

Approach: *Buy/Puzzle*. This is the strategy to be used in soliciting funds for basic research. The procedure is to demonstrate how the particular puzzle lies at the centre of a nest of concepts which in turn help us to understand certain phenomena. It will obviously help your cause if these phenomena can in turn be shown to underlie the solution of a number of practical problems.

Strategy L: "There are warning signs that..."

Approach: *Modify/Puzzle.* The basic message in this strategy is likely to be: "If we find x then we will need to do y". It is a strategy to invoke action to insure against difficulties that may arise if certain current researches into fundamental issues come out in specified directions.

Strategy M: "Why theory x is better than theory y..."

Approach: *Research/Comparison.* This strategy is the classic "conflict of views" approach in which rival theories are appraised for their capacity to deal with certain data. It can be deployed in a wide variety of contexts from university lectures (where the object may be to show why new ideas have supplanted old ones) to presentations to funding agencies (justifying researches which will provide definitive evidence to support a new theory against an old one).

Strategy N: 'The 'Which?' survey...'

Approach: *Buy/Comparison:* I have called this strategy the "*Which?*" survey because it is deployed in classic manner in the journal of the Consumers Association '*Which?*'. Products are systematically compared against a number of criteria with a "Best Buy" section to assist those who have difficulty in deciding between possibly conflicting criteria.

Strategy O: "Our output will be greater if..."

Approach: *Modify/Comparison:* This strategy, like many of those reviewed, is another version of the thesis. In this strategy, the technique is to compare one procedure with another, demonstrating step-by-step that some changes in practice will yield specified benefits.

Strategy P: "These explanations won't do because..."

Approach: *Research/Ninepins.* In seeking to justify further research, a more adventurous procedure than a straight comparison of theories is the approach of systematically demolishing all but the favoured theory. If you enjoy a combative style of argument, or if you are comfortably on "home ground", this strategy can add spice to a presentation.

Strategy Q: "See what rubbish this is..."

Approach: *Buy/Ninepins.* If you find the "Which?" approach too tame, you may prefer to demonstrate the singular virtues of your system, device, or procedure by showing the inadequacies of those of competitors. Destructive testing of actual exhibits can be a vivid way of illustrating such a strategy.

Strategy R: "These systems don't meet our needs because..."

Approach: *Modify/Ninepins.* Again, the "Ninepins" approach can be a provocative way of urging modifications. It is an approach to be handled with some care, because if you use it in the presence of people whose ideas you are "rubbishing", you can expect vigorous resistance. On the other hand, if you enjoy conflict... .

Chapter 4

Starting, Stopping and Signposting

You, like me, have probably felt frustrated when a speaker has launched into a presentation, perhaps pouring out facts and figures which may or may not be important for us to remember, without any indication of what is coming. At the very least, we are left not knowing what we might wish to note down; at worst, the speaker may fail to establish some important idea or fact because we have not been offered an intellectual framework through which to perceive it and retain it.

The message of Chapter 3 of this booklet was to select a structure that gives appropriate emphasis to your ideas; the message of this chapter is to make sure that your listeners know what that structure is so that they can give full attention where it is needed. They need to be able to discern the figure against the ground, to distinguish the key ideas and information from the background that throws them into relief. The way to achieve this sharpness of focus is to

> **Make sure your listeners know where they are at all times.**

Use the start of the talk to give a map, provide map references as you go along and signal when you reach your destination.

When we read a technical report or scientific paper, we can move backwards and forwards through it not only to remind ourselves about points, but also to home in on the bits that interest us. Having identified a promising paper through an Abstract (summary), many of us read a technical paper in an order different to that in which it is presented. For example, we may read the references first to see if the paper is indeed in the "right ball-park". Then we may read the conclusions, followed by the summary of principal findings. If we doubt the validity of the findings, we may then read about the procedures used, although they will probably have been described before the findings are listed. What we need to remember is that

> **A person listening to a presentation cannot move backwards and forwards through the content in the way that someone reading a paper can. For this reason, it is essential to provide strong framing at the beginning of a presentation.**

Starting

The moment when you are in front of an audience, on your own, ready to begin, all eyes on you, is without doubt the most stressful of all. There is often a temptation to try to pretend that the moment has not actually arrived — by fiddling with things, waffling, or by uttering platitudes. Don't!

At the very beginning of your presentation, you have a unique opportunity to grasp your listeners' attention. You will spoil the opportunity if you: tap or blow the microphone or twiddle its stand; look down at your notes; shuffle overhead projector transparencies about on the table; complain about the traffic jam that nearly prevented you getting there on time; and so on and so forth. Make sure that all "stage properties" are in place and working before you are introduced or make your formal start (see the check list in Appendix A below). Then

> **Don't waste time on preliminaries. Go straight into your presentation.**

Avoid apologies, elaborate statements of thanks for being invited to speak, or any other clutter. (You can always say these things at the end if they really matter.)

> **State your subject.**

At a large conference, you may wish to have a slide up as people come in, giving the full title of your presentation, your name, and your institutional affiliation. With the programme changes that happen in any major conference, it helps if people know what your presentation will be about before you start speaking so that if they are in the wrong room, they can go quietly without disrupting your opening.

State your purpose.

Recalling the maxims offered in Chapters 1 and 2, this should always be in terms of what your listeners need to know, not what you want to say. Explain what you hope they will do as a result of hearing your presentation — even if this is only to read a specific book or paper. More probably, you will have a more complex message to convey and may be using one of the strategies outlined in Chapter 3 above. On the assumption that, as a professional, you are not trying to subvert choice but rather to stimulate choice, tell your listeners what your strategy is and why you think they will benefit by doing what you will be suggesting.

State your plan.

This needs to identify important points that you will be making in, for example, supporting a thesis. If your listeners know in advance what the key points will be, they will be better prepared to attend to them as you reach them in the course of your argument, and they will not be distracted by less important details. You are, in effect, offering a sort of "cognitive map" and identifying landmarks.

An analogy may help. If you were a stranger to London and asked me how to get from Imperial College to Tottenham Court Road by bus, I could reply simply: "You take a Number 10". This is true — but less than helpful in ensuring an interesting and memorable journey. If, instead, I were to say: "You take a Number 10 bus going East from the North side of Kensington Gore. The journey will probably take about 35 minutes. You will pass three major landmarks: Hyde Park Corner (where the bus will turn

North), Marble Arch (where the bus will turn East to traverse the entire length of Oxford Street), and finally Centre Point (where the bus will turn sharp left to go North again). Get off when it makes this last turn." These additional "direction-finding" comments help you to enjoy the journey so that instead of twiddling your *"A to Z"* and wondering where you are, you can look out of the window and enjoy the passing scene. By picking out a few important points, I put you in a position to take in many more. But note too: if I give you too many directions, you will still be left struggling with your *"A to Z"*!

In a technical presentation, your listeners need one or two key ideas to look and listen for; they can then attend as they choose to the subsidiary points through which you develop the key ideas. In starting a presentation, therefore,

> **Identify the strategy, the line of argument, the theme, the plan, as soon as possible.**

If your listeners are primed at the start to appraise your proposed system, device, or procedure in terms of its economy, elegance, fertility, reliability, ingenuity, they will be in a position to judge whether or not the details you offer are sufficient for their purpose. They are more likely to remember what you say than if they have to wait to discover whether or not specific points are important.

Having attracted your listeners' attention, you need to ensure their interest. You can do this if you

> **Relate your material to your listeners' experience.**

One of the most powerful ways of doing this is to give your listeners an actual experience of what it is that you propose to explain or analyse in your presentation. This is why it is valuable to

> **Demonstrate an effect, show a film, or activate an exhibit at the beginning of a presentation rather than at the end.**

Another way of linking your ideas to the experience of your listeners is to tell a story. Stories are usually built into the "Disaster" strategy of organisation; but many other forms of organisation come to life if, after establishing the framework of ideas, you address the questions which a good news story answers: who? what? when? where? why? The detailed substance of your presentation can then answer the remaining question: how?

To decide upon a suitable "story" with which to earth your ideas within a context of your listeners' experience, consider:

- **topical points** (recent news events, bad weather, etc.);

- **local colour or details** — which will incidentally demonstrate that you have taken the trouble to find out something about your intended listeners;

- events from **your own experience** — which can also have the useful additional effect of demonstrating your credentials to talk about the subject.

Common experiences, whether actual or vicarious, draw people together and create a receptive atmosphere for a presentation. However,

> **Avoid openings which alienate, distress, bore, or in any way irritate your listeners.**

Such openings to avoid include:

History — a long recitation of names and dates, unless the development of the system, idea, procedure, device really is relevant to your presentation.

Humour — unless you have a very sure touch! The ritual joke to "soften people up" at the start of a presentation is usually perceived for what it is — time-wasting before you get to the serious business. The best jokes are, as you know, culture-specific, celebrating and reinforcing the beliefs (prejudices?) of specific social groups. If you misjudge your audience and/or alienate some of your listeners, you can dig yourself a pit from which it is remarkably difficult to escape.

Startling openings are sometimes recommended to arrest the attention of your listeners. Handle with caution — unless you have St. John's Ambulance officers in attendance. I have heard a lecture opened by an explosion: great stuff — except that several members of the audience spent half of the lecture cowering behind seats with their fingers in their ears, and others were so anxious about the possibility of another explosion that they missed half of what the speaker was saying. Amazing facts and figures can have much the same effect: they certainly can capture our interest and attention — but, unless you are careful, they may do so at the expense of a deeper attention to the structure of ideas.

Likewise, **controversy.** Don't stir up hornets unless you positively enjoy swatting them! You may, however, have been asked to speak precisely because the topic is controversial: there are special techniques for handling lively situations — see the section on public inquiries in Chapter 8.

Signposting: Emphasis of important points

Having given a map or plan at the beginning of your presentation, stick to it. Faced with rows of staring faces and experiencing none of the feedback that helps you in a one-to-one conversation, you may feel tempted to add in comments to embellish what you had planned to say. Resist the temptation!

> **Don't depart from the plan in your notes.**

If you have thought your presentation through carefully beforehand in terms of what your listeners need to know, you should not need to deviate from your plan. This is why it is so important to learn as much as you can about your audience before you plan the presentation.

You will find that a good structure helps to keep your thoughts in order, and so will your listeners. One way to keep on track is to

> **Phrase the headings in your presentation as questions.**

In a conversation, your remarks are automatically structured by the questions you are asked. When you are speaking without the benefit of direct questions, you can make the subject more accessible to your listeners by articulating the questions you think they will wish to have answered at each point in your presentation. By building around questions, (which you can highlight in your opening remarks), you are providing the signposting necessary for your listeners to know where they are.

You can help greatly if you

> **Summarise frequently.**

Re-state the main idea or point of each major section before going on to the next section. This simple procedure will do wonders for your pacing.

Just as you need to let listeners know where they are within your structure at all times, so it is important not to leave them wondering what matters most. Point up the significance of particular points by specific mention:

(Pause. Look directly at the audience.): "This is terribly important!"

Stopping

The end of a presentation is even more important than the beginning. When you signal that you are about to stop, people will perk up — perhaps even seizing their pens to note down your final distillation of wisdom.

Getting off stage can be a problem in some forms of public presentation. Even Shakespeare, desperate to find a way of removing Antigonus in Act 3 Scene 2 of *"A Winter's Tale"*, has the rather feeble stage-direction "Exit pursued by a bear"! In speaking technically, you need no such expedient. The message is,

> **Always summarise at the end of a presentation.**

Re-state the main point (which presupposes that there is one!). If, as I recommend, you build your presentation around action which you hope your listeners will take, re-state in the summary what that action should be.

Always be precise about what immediate action your listeners can or should take.

Where the intended action is Research, you may wish to finish by exhibiting a slide giving the full reference to a relevant paper, plus your address and e-mail.

Where the intended action is Buy, give necessary details of the Product (such as cost) plus, again, the address and e-mail of someone to contact.

Where the intended action is Modify, specify precisely what action should be undertaken by whom, by when, and what it will cost. Again, leave your listeners with an address and e-mail.

The summary should pull everything together crisply. Therefore,

> **Do not put new material into the summary.**

Even if you think you have under-shot your allotted time, do not burble on with extras at the end. When you have finished what you planned to say, sum up and stop. If (which is most unlikely) you do use less time than you had planned to use, say cheerfully and confidently: "I have left time for questions" — but resist the temptation to keep talking for the sake of it.

If, as you are speaking, points occur to you which you think your listeners should know, do not divert from your plan, but rather save them until question time. If no-one asks a question, you can easily add in a comment at the end of the proceedings.

Chapter 5

Speaking Technique

The essence of good speaking technique is your frame of mind. If you are prone to feel nervous, remind yourself that

> **The object of speaking technically is to keep your listeners' attention on your subject, not on you.**

In this regard, speaking technically may well differ from political speaking. If you protest that sometimes the object of speaking technically actually is to draw attention to yourself (for a job, in competition for research funds, etc.), I would argue that a technical presentation which conveys content effectively will indeed draw attention to you — as someone professionally competent, who thinks and speaks in an orderly manner.

Think about your listeners, not yourself! Think what they need to know — not what you want to say. Not only does this orientation help you to plan the content of your talk, it also helps to put you into the correct frame of mind.

You are usually speaking because you have been asked to. That is to say, you are there because you are acknowledged to know your subject. (I deal in Chapter 9 with the technique for handling a question to which you do not know the answer!) Furthermore, by having read thus far, you have studied some of the most important techniques for speaking technically. Any remaining nervousness will diminish if you remind yourself that

> **If you have thought through your material carefully in terms of what your listeners need to know, rather than what you want to say, any symptoms of nervousness in your speaking technique will probably not be noticed by your listeners.**

One of the reasons that I use video (CCTV) in Speaking Technically workshops is that people find this proposition remarkably difficult to believe — until they have seen themselves on the screen! Butterflies in the stomach cannot be seen, and the techniques listed below can remove most of the other symptoms of nervousness.

To get yourself into the right frame of mind,

> **Concentrate on things to do rather than on things not to do.**

The crucial "do" is to have a coherent plan and stick to it. There are a few "don'ts" (see below); but they are far less important than being well-organised. You will feel less nervous if you have planned properly.

You will also feel less nervous if you know the symptoms of nervousness and how to control them. For this purpose,

---

Be aware of physical symptoms which are normal.

Certain symptoms are produced by adrenaline starting to flow before it can be put to use:

- faster heart-beat;

- "butterflies" in the stomach;

- dry mouth or a need to swallow;

- knees or hands shaking.

You will already have noticed that these symptoms disappear quickly as soon as you start speaking. The best approach is, therefore, to ignore the symptoms of nervousness.

The following matters do, however, merit attention:

Audibility: Speak more emphatically than in normal speech.

A few ways to do this:

- Articulate your consonants;

- Breathe deeply;

- Exercise your jaw and open your mouth, for example by practising saying (in some secret place!): "Tarara boom dee-ay".

> **Quality: Decide what matters most — your voice will adjust.**

If you have planned your talk carefully, you are bound to want to emphasise some points more vigorously than others. Rather than going for elocution lessons, let the emphasis of your thinking determine the emphasis in your speaking.

If, however, you still sound monotonous, try practising a sentence with different types of emphasis. For example, say several times "I opened the box and it was empty" imagining each time that you are in a different frame of mind — angry, apprehensive, confident, happy, indignant, nervous, pleased, tired, terrified, etc.

If you listen to yourself on a sound tape-recording without vision (a practice to be avoided, I suggest), you may notice the odd "er" and "um". Very rarely, in my experience, do listeners whose interest in the content of a talk has been captured notice "ers" and "ums". If you, and anyone you persuade to listen to a practice presentation, do notice excessive "ers" and "ums" — cure yourself by planning pauses for thought in your presentation. Mark these in your notes: (see below).

If you listen to yourself on tape, you may worry about your accent. Don't! As long as you articulate clearly, no-one else will worry about your accent if you don't — provided that what you have to say is interesting.

> **Speed of speaking: Mark notes with timing points.**

When we are nervous, we tend to speak more rapidly than usual. As I have indicated above (Chapter 2), the key matter to watch is pacing: the number of words-per-minute is much less important than the number of ideas-per-unit-of-time. But the two are obviously related. If you start to gabble through your material, your pacing may go to pot. So, mark your notes with timing points, and pause for breath in appropriate places — for example after each section-summary.

> **Manner and manners: Think of your presentation as a conversation.**

There are two sides to this axiom — intellectual and social. I have already recommended (Chapter 4) that you think through the questions that your listeners may have. These questions can provide the intellectual framework of your presentation, and even act as sub-headings in your notes.

The notion of a conversation is also important in establishing an appropriate social relationship with your listeners. Speakers who appear hectoring, surly, remote, arrogant, or patronising probably do so because they have not thought through their material from their listeners' point of view. They would probably be quite agreeable in conversation but come over badly because they are ill at ease in the more formal context of a presentation.

To think of a presentation as a conversation (albeit one in which you both ask and answer the questions) can be a guide to what to wear. Clothes convey important non-verbal signals. For a conversation, you would wear clothes appropriate to the specific social situation: let the same principle guide you for the more formal context of presentations. To wear white tie and tails at an informal postgraduate seminar would be as inappropriate as to wear an open-necked shirt and jeans at a Royal Institution Friday evening lecture.

> **If in doubt about what to wear, ask your hosts. If you cannot get clear advice, err on the side of formality.**

It is much easier to dress down (take off a jacket) than to dress back up.

> **Posture and movement: Stand with feet slightly apart, heels feeling the ground.**

This way of standing looks good; it also feels good — probably because it is the position in which we would stand if preparing to resist an attack! More importantly, the simple trick of letting your heels feel the ground not only prevents your knees knocking, but also helps to keep your back straight but not stiff, and helps you to avoid slouching, drooping, leaning over the table or overhead projector, or in other ways looking messy.

As in a conversation,

> **Your aim should be to make your listeners feel comfortable.**

You will succeed in doing this if you avoid doing things which make them feel uncomfortable. So, if you need to move about, do so, but don't pace about unnecessarily. Try not to fidget: put paper clips, pencils, chalk, etc. out of reach. Check your clothes before you go in front of your audience so that you do not need to hitch them up or fiddle with them (tie, buttons, zips, cuffs, etc.). If you find your hands are always drawn to your pockets, sew up the pockets on your conference suit to prevent yourself putting your hands in them. (The habit of jangling coins, particularly if performed by rich industrialists in university seminars, drives impoverished academics and students mad!)

If you do not need the pointer for slides, put it out of reach: the temptation to lean on it, slash the air with it, twirl it like a drum-major's mace, etc. is almost irresistible!

Gestures: Be natural.

As with all aspects of technique, don't worry about what not to do; decide what you will do.

When you first give presentations, you may find that your hands feel big. Don't worry; they aren't! But to make yourself feel better, have something to hold — such as your notes (see below).

As you get used to planning presentations, to achieve emphasis you will almost certainly start to use your hands as you do in conversation. You can in fact help your listeners greatly if, for example, you count points on your fingers hold hands out to show 'this way' or 'that way', or to indicate size or height.

Don't, however, thump the table for emphasis — except perhaps once a year (when the teacups should bounce)!

Look at your audience.

Imagine a conversation in which the person you were talking with did not look at you. It is essential to establish and maintain good eye contact.

If you find the prospect of people looking at you too appalling, use slides. They will then look at the slides, not at you. But you still need to look at your listeners to see if they are following.

To give yourself the opportunity to look at your listeners, you need to have notes in an appropriate form — so that you do not need to spend your time reading them.

Planning your notes: Object — to help you to think on your feet.

Always have notes — even for a so-called "informal" talk. Rather than thinking that you do not know your stuff, your listeners are likely to feel gratified that you have clearly taken the trouble to plan what you are going to say.

A good set of notes is the best insurance against nervousness.

It does not help to have everything written out — because written prose is more dense and formal than speech. Unless you are exceptionally good at writing words to be read out, you are likely to sound rigid — and you will almost certainly be hard to follow. You will also find it hard to maintain eye contact.

In short,

> **Do not read from a script.**

Do not read unless the facts have to be absolutely right (as at a public inquiry, or in a court of law — where you may need to read out a description of an event, sums of money, clauses of a contract). Even in a highly formal occasion, usually only part of your overall presentation may need to be read out. You may often be able to incorporate key facts on a slide so that everyone can see them.

> **Do not memorise a presentation.**

Not only will you sound as if you are reading (because you are likely to have learned a "script"), but you will also build anxiety — worrying lest you forget something. If you memorise a text, you will find it hard to make changes (e.g. to adjust to what a previous speaker has said).

> **Plan your notes so that you can think on your feet.**

Divide your material into small sections so that, for example, you are giving not a 20-minute talk but rather 20 one-minute talks. In short, build around a series of interlocked headings and sub-headings on each of which you give an "impromptu" one-minute presentation.

In planning a presentation, particularly if you do not have much experience, you may wish to write out a complete text — mainly to determine how many words you think you might need and to check this against the time you will be allowed. Allow 100 words per minute — even though you think you might speak more quickly than this. That is to say,

> **In estimating the length of your presentation, always aim to under-shoot your allotted time.**

Very rarely will you in fact stop short of the allotted time; but if you do, you can take more questions!

Make notes from your text, and then THROW IT AWAY! The notes should be in the form of headlines on which you can elaborate.

When you feel more confident (or, I would argue, to *make* yourself feel more confident),

> **Start with a plan of headings and sub-headings in your notes and then expand.**

Lay the notes out so that you can read them at a glance. Unless you have very good eyesight, you will probably find type too small for this purpose. You will find it helpful to vary the size of writing for your major and minor headings, indent them hierarchically, and colour code them for emphasis.

Your notes might then look something like this:

> **Jumbo-Sized Major Headings** (in red)
>
> Indented Minor Headings (in blue)
>
> Key data, formulae, etc. further indented (in green)
>
> Comments (in black) e.g. Switch on OHP, timing markers.

You may be lucky enough to have a lectern, but do not bank on it. If you have only a table, this will almost certainly be too low — and you will be tempted to stoop over it. So

> **Always assume that you will have to hold your notes in your hand.**

On some occasions (very short presentations, after-dinner speeches), cards may be useful. If you use them, write on one side only; number them; punch a hole in the top left corner and tie them together with a treasury tag; do not write over the entire surface, but leave space at the bottom for your thumb.

On most occasions, however, you will probably find a clip-board more useful. If you write your headings on an A4 sheet, you may be able to put the notes for the entire presentation onto a single side. (If you cannot do this, you might ask yourself if you may not be going into too much detail — or giving people more than they can reasonably be expected to take in). With writing of decent size, you will be able to see the structure of your talk at a glance. Sheets of A4 are, however, difficult to manipulate: they droop, or, worse still, magnify any residual tremor in your hand! A clip-board looks business-like: it also keeps things under control.

In many situations,

> **You may find it useful to use slides both to help your listeners and to serve you as notes.**

If you do this, it is crucial not to make the slides too complicated: see Chapter 6 below.

Chapter 6

The Use and Abuse
of Visual Aids

This chapter is about the use (and abuse) of visual aids; it is not about their production. Most organisations have specialists to make slides, films, and video-tapes for presentations; often organisations expect materials to be made in a "house-style", perhaps containing a company logo on each slide.

> **Plan the visual aids as soon as possible so that you can arrange for them to be properly produced.**

Your main decision will be the strategic one of what sort of visual aids to use. To this end, you need to

> **Find out before you plan the presentation what facilities there will be in the room in which it will be given.**

If, for example, there is no black-out, you will not be able to use film or 35mm slides. If many people are expected, you will be able to use video only if there are multiple monitors (roughly one for every 20 people) or if there is a special video-projection device.

Also, at the planning stage (rather than on your way to give a presentation),

> **Think what you will do if the projector/computer breaks down!**

You will also need to consider whether the occasion merits the cost of producing certain types of visual aid. The following list highlights some of the issues:

	OHP	35mm	Video	Film	Mmp
Cost	Low	Modest	Moderate	High	Moderate
Preparation	Easy	Moderate	Moderate	Difficult	Moderate
Flexibility	Easy to adjust	Hard to change	Easy to "search"	Very inflexible	Easy to "search"
Theatricality	Low	Moderate	Moderate	High	Moderate
Lighting	Normal	Dimmed	Normal	Dimmed	Dimmed

The basic rule to follow is

> **Only use visual aids if they positively reinforce what you are saying: never use them for mere decoration.**

The ease with which overhead projector (OHP) slides and/or direct computer output may be produced with a personal computer and Multi-Media Projector (Mmp) offers you a grave temptation to do fancy things. Resist it!

The observations which follow will help you to decide when and how to use visual aids; they give suggestions which will help you to brief your graphics department.

> **A very rough rule-of-thumb is always to use lettering or numbers on an OHP slide or Multi-Media Projector Slide of 24-point or above.**

Visual aids can be both helpful and damaging — sometimes both helpful and damaging within a single presentation. That is why the Check Sheet in Appendix C invites observers to add marks for a positive use of visual aids and subtract marks for a damaging or inappropriate use of them.

Visual aids can emphasise important points, help to pace material; and add interest to a presentation. They can, however, distract attention from what you are saying, slow a presentation down too much or (more frequently), bewilder the audience with too much information. The key point to keep in mind is that

> **We cannot look at a visual aid illustrating one thing and simultaneously listen to you saying something else.**

You may have had the misfortune to suffer from this phenomenon if a speaker has flashed up a chunk of text and gone on talking while you have dithered between trying to read the text and listening to what the speaker was saying; or become absorbed with handling an exhibit and missed everything that was being said about it; or struggled to read the axes of a graph while the speaker was rushing into an account of what the graph purported to show.

> **The proper use of visual aids is for emphasis and highlighting: do not attempt to show everything.**

Just as the presentation as a whole should not attempt to say everything, so the visual aids should not attempt to show everything. If you pile in too much, your listeners will just become confused and exhausted and will remember nothing. That is why it is essential to

> **Link the visual aids you use in your presentation with the documentation you use to support it. See Chapter 7 below.**

If you follow these few basic principles, you will have no problem in planning your use of visual aids. The following notes simply pick out a few points for emphasis.

(a) Overhead Projector Transparencies

- Keep as simple as possible;
- Beware of too much material: if people need to study a transparency at length, it is probably too complex;
- Beware of too much detail:

Use round numbers and avoid decimal points;
Write 1 in 10 rather than 10%;
In tables, limit yourself to a maximum of five rows
and five columns;

- Do not go too fast or use too many transparencies;

- Do not include any word or number not worth mentioning: but mention every word and number that you do include;

- Point to and mention key features, e.g. axes on graphs: do not leave people to figure things out for themselves while you go on talking;

- Rather than include many details on one transparency, use overlays (i.e. putting one transparency on top of another) to add material in;

- Try to avoid "cover-and-show" (i.e. pulling a piece of paper down a transparency to progressively reveal material): some people find this technique intensely irritating — because they feel patronised (as if the speaker thinks that they could not decide for themselves whether or not to read ahead of what the speaker is saying);

- For informal occasions, consider the possibility of using "skeleton slides" done with permanent (waterproof) ink to which you can add (or from which you can wipe off) material written with non-permanent (water-soluble) ink;

- Use colour coding (as in your personal notes) to signal major headings, minor headings, details and so forth;

- Do not switch the projector on too soon; do not leave it on when it is not needed;

- Say what a slide will show before you put it on; pause when you put it on; then take people through it.

(b) 35mm and Multi-Media Projector slides

Most points are the same as for Overhead Projector Transparencies with these additional points:

- Rather than include many details on one slide, build a diagram/table/list of points by showing a sequence of slides with the new material highlighted or in bold lettering:

The first slide will show one line.

The first slide will show one line.

The second slide adds a second line.

The first slide will show one line.

The second slide adds a second line.

The third slide adds a third line.

- Try to bunch 35mm slides in one part of the talk so that you do not have to keep raising and lowering the lights;

- If this is not feasible (with a bit of planning it should be!), rather than leave distracting material on the screen, use blank (black) slides to break a sequence;

- Switch the projector on before your presentation, using a special focussing slide to achieve a good picture: use a blank (black) slide so that you can have the projector running ready for when you need it;

- Never ask a projectionist to re-show an earlier slide: use duplicates instead;

- Use only relevant slides: resist the temptation to show slides just because you have them!

- Beware slides with irrelevant detail (e.g. scenery or buildings behind a piece of equipment); plan photographs carefully and/ or ask your graphics people to modify the slides.

(c) Blackboard/Whiteboard

Although most technical presentations will be planned in advance, and merit carefully planned and properly produced visual aids, a blackboard (with chalk) or a whiteboard (with spirit pens) can be very useful for illustrating points in answer to questions.

With very complex material, the discipline of writing can slow the speaker down to a speed appropriate to the capacity-for-understanding of the listeners. Indeed, it is not unknown for university departments to forbid the use of the overhead projector to lecturers not on grounds of cost, but because the students find that the lecturers go too fast if they do not write things down as they go along! I do, however, suspect that in these circumstances, lecturing is being used for inappropriate purposes — for the transmission of information (which is better left to books and computer programmes) rather than for the enhancement of understanding which the time-lapse-photography, map-making, helicopter-viewing of speaking technically can do superbly.

The "rules" of blackboard/whiteboard work are pretty obvious:

- Look at your audience as often as possible;
- Do not stand in front of your writing;
- Plan your lay-out (even when you do the writing impromptu);
- Write big;
- Beware of mathematics: key in with paperwork so that people are not torn between trying to get a correct record and understanding what you are saying (see also Chapter 7 below).

(d) Flip-charts

Flip-charts are pads of large (usually A3 size or larger) sheets of paper on a stand. They can only conveniently be used for very small audiences — for example in an office or boardroom. Their advantage is that they are portable and free you from dependence on projector bulbs, screens, cables, computer Gremlins and other hazards.

Regarding the organisation of information, the same basic principles apply as with overhead projector transparencies and 35mm slides. Specific procedures are to:

- Keep material covered until it is needed, and cover it when you have finished with it: this can be achieved by interleaving blank sheets between the illustrated ones;

- If you intend to write or draw during your presentation, draw lightly in pencil in advance and then trace over with the felt pen.

(e) Demonstrations/Exhibits

If you have a "bag of tricks" to demonstrate, show it at the *beginning* of your presentation — and then again *during* the presentation to illustrate particular points. By showing it at the beginning, you not only catch your listeners' interest, you can also plant questions (in one of the ways discussed in Chapter 3 above) to provide a thematic structure to your presentation.

A film or video-tape can serve much the same purpose. This can be particularly valuable as a prelude to a site visit — so that visitors can get an over-view before being confronted with details.

Other comments are fairly obvious:

- Exhibits must be big enough to be seen — unless you are specifically making a point about how small something is. If we need to see detail, it may be better to show a blow-up diagram than to show the actual object.

- If possible, give your audience something to do — count, estimate, listen, hold, etc. The theatricality of exhibits merits on some occasions a certain theatricality of approach.

- Never pass an exhibit around. Apart from the danger that you may never get it back, you run the risk of losing the attention of the person(s) actually looking at the object, and irritating the people down the line who do not have the object to hand at the moment when you are talking about it. By the time these people do get the object, you are likely to be onto some other part of your presentation.

Chapter 7

Planning the Support Work

This is a brief chapter, but an important one. It is written separately to give due emphasis to an often overlooked proposition, namely that

> **Technical presentations are rarely isolated occasions: meetings, actions, discussions, questions, and paperwork must be planned within a single, overall strategy.**

Sometimes when technical presentations have been unsuccessful, the speaker's fault is that of putting too great a burden on speaking technically — and failing to see the speaking as but one of a wider strategy of exposing the target audience to ideas and information.

On many occasions, speaking technically has an almost ceremonial function. For example, the main purpose of a presentation may be to welcome guests before they go off with fellow-specialists in different sub-areas of work for detailed question-and-answer sessions.

Again, speaking technically in the form of lecturing may be primarily a form of advice rather than the transmission of information. A system of instruction at a university may consist of Materials, Advice, and Assessment. Although attention may sometimes focus on the performance of lecturers, the three components need to be considered together. In principle (and often in practice), students can find their way through materials on their own with the stimulus and focus of assessment; in a very real sense, teaching is a form of advice which helps them to use the materials more efficiently and effectively in preparing themselves for assessment.

> **It is an abuse of the possibilities of lecturing to make it the primary vehicle for conveying information.**

If students cannot discover the syllabus (i.e. that on which they are to be assessed) without attending the lectures, enormous opportunities for flexible education are lost.

Likewise, many forms of speaking technically (such as conference and industrial presentations) are initial moves in a complex set of interactions between interested parties. These interactions can be seriously impaired if you take up too much time in talking — and if you do not plan the speaking in conjunction with paperwork sent to people before you meet them. The presentation may, on such occasions, be primarily for highlighting detail in what people have already read so that they are in a better position to ask questions or take part in discussions.

> **Paperwork distributed before your presentation can prepare your listeners.**

It can do so by:

- offering an initial overview of your subject;

- giving full details so that they can plan questions;

- help them to follow your presentation by incorporating key diagrams, formulae, performance statistics, and other such details for record.

The cost of assembling senior people for a meeting of any sort, including a technical presentation, is considerable. Try a modest calculation for the next meeting you go to.

For example, the Civil Service undertakes efficiency studies on the assumption of a working year of 213 days. Calculate full-employment costs on a day-rate basis using this figure. Add in an appropriate figure for overheads (this may well be over 100% of professional salaries if done realistically), travel costs, catering, and overnight accommodation — and you soon begin to realise what "ball park" you are in. Compared with these costs, the costs of distributing preliminary paperwork are minimal.

For many meetings of professional people, well-planned preliminary paperwork is an essential element of efficient and effective working. But, even when you can rely on colleagues doing their homework,

> **Do not assume that people will have absorbed all the details of the paperwork: plan the presentation and the preliminary paperwork around a single set of interlocking headlines.**

Where presentations are part of a course of instruction, students find it very helpful to have a detailed outline of material to hand not only so that they can find their way through the ideas, but also so that they can plan their work — and insure against enforced absences. If you think preparatory paperwork would make them less inclined to come to your lectures, you might think what the lectures are really for. What, in short, is the added value of attending a lecture over and above receiving the material in written form only? If the lectures serve their proper functions (of map-making, advice-giving, helicopter-viewing, and so forth), they will be complementary to the notes in important ways which the students will be quick to perceive.

> **Whenever it is important that people should have accurate information, do not rely on a spoken presentation alone to give it, but offer it in printed form too.**

On most occasions, there is much merit in sending such information before people come. If they know in advance that they will have a correct record of important data, they can attend to your argument without worrying about whether or not their notes are full and accurate. Certainly if you use one of the strategies outlined in Chapter 3 above, your listeners will have the stimulus of judging whether or not you draw on the data effectively to 'make your case'.

> **Do not table papers during meetings.**

A counsel of perfection, I know! Many meetings have an urgency that makes it necessary to provide material only as people arrive, but many meetings do not! I have seen conference presentations ruined when speakers hand out a detailed paper at the beginning of a session and then hop about its contents leaving their listeners in frustrated confusion.

Sometimes it may be both necessary and desirable to give people material for use after the presentation — for example, to consolidate what you have said; to ensure an accurate and permanent record of key information; for use in subsequent discussions.

> **If you are going to give out material after your presentation, say so at the beginning.**

You will thereby relieve your listeners of anxiety about taking notes and/or trying to remember facts and figures. In my judgement, however, people benefit more from practically any presentation by knowing what to expect than by being left in ignorance of what they will be offered to take away. Paperwork sent out before a presentation is usually more effective than paperwork made available afterwards.

Whatever you decide to do,

> **Plan the total operation — paperwork, ancillary meetings, presentation(s) — in terms of the action your listeners will take.**

As I have indicated in Chapters 2 and 3 above, a strategy focusing on the action you want your listeners to take can give a cutting edge to many types of technical presentation. The type of action to be promoted depends on many things — not least the type of occasion in which you are speaking technically. The following chapter, therefore, offers some notes on the purposes of some specific occasions. Again many of the points are self-evident, but the checklist may be useful.

Chapter 8

Some Special Situations

Your strategy in speaking technically, in terms of choice of material, theme, and approach will reflect your interpretation of your audience's interests and needs. The structures discussed in Chapter 3 offer help at the strategic level for most situations. There remain, however, some tactical issues to be considered concerning the type of expectations generated by, and the conventions usual in, some specific situations in which you will give presentations.

In all cases,

> **Decide the fundamental purpose of the occasion.**

There is an almost infinite variety of purposes to be served by technical presentations; the following brief notes do not, therefore, attempt to be comprehensive. Rather, they offer some observations on typical purposes of some situations (with associated comments on practicalities) which will help you to fine-tune your presentations (situations listed alphabetically).

After dinner speeches: Celebrate the group.

Very frequently, after-dinner speeches have the ceremonial function of celebrating membership of an organisation. The dinner itself is the ritual by which membership is defined and enjoyed; the speech afterwards offers intellectual lubrication to a basically social process.

The popularity of jokes and stories on such occasions reflects the desire for belonging. The problem is that the best jokes rely for their effect on the similarity of experience of those who enjoy them. In closed social groups, even very modest stimuli will give rise to much mirth; but in public situations, jokes are very difficult to tell well. Lowest-common-denominator jokes are simply embarrassing in mixed company.

You may prefer to concentrate on episodes of collective memory — recollections of the achievements of the group and/or of specific members of it. Done with lightness of touch, this can have much the same effect as joking in generating a feeling of belonging.

Alternatively, Prospect may be more appealing than Retrospect — a foretaste of how the group may cope with coming problems: changes in legislation; further cuts in funding; increased competition; growth in membership; and so forth.

In short, it may be best to leave the joking to consenting adults in private and concentrate on something of substance.

Conference presentations: Start a conversation.

Most people value conferences for the opportunity for detailed discussion with other workers in the field. While results may be presented for the first time in conference, it is more likely that those present will at least have heard rumours about your work even if they have not read about it. They are almost cetainly more interested to interact with you than to hear you read out something which they could equally well read for themselves. Do not, therefore, literally read out your paper. Think of the spoken version of your paper separately from the written version.

If you have the opportunity to present an abstract for circulation before the conference, try to make the abstract a coherent, self-contained summary of the work you wish to discuss.

Describe succinctly what was done, by whom, when, where, why, how (with summary points from the findings), with what effect and (if possible) what it cost and who paid.

In the presentation, highlight issues upon which you would welcome comment and discussion. Ask for help and advice in solving problems and puzzles rather than boast of your achievements (see strategies in Chapter 3). This way you will reduce antagonism — and maybe start a research collaboration.

Be brief. An allocation of maximum time for your session and a suggestion of minutes for your opening remark will probably be in the conference plan. If you can use less time for your opening remarks than indicated, so much the better: there will be more time for discussion.

Point up as clearly as possible what future action should be taken by whom (research, good practice, etc.). This itself will probably stimulate discussion.

> ## Coping with visitors: Don't try to show them everything.

A very common type of speaking technically, deceptively simple because of its apparent informality, is the presentation given to visitors to a laboratory or other site. Such meetings can prove very fruitful, but they can be frustrating — particularly if, through not being well-organised, you try to tell and show too much.

When you are asked to receive visitors, find out who they are and what are their interests. Do not have a standard presentation for all-comers, even if you have some favourite engines to wheel out! Keep to time or visits will get into chaos; (in case the person ahead of you in the visit has not read this booklet, you may need to plan to use less time than the schedule offers — to allow for visitors being late). Have notes and do not waffle.

Check if translation will take place; this will double the time needed. If there is an interpreter, speak to him/her not to your visitors; look at your visitors while the interpreter offers the translation.

If you will be showing visitors round a site, have a briefing session before you set off round the site. If your site is noisy, cold, or in any other way inconvenient, have this session somewhere reasonably quiet and comfortable (see also Industrial tourism below). Decide where to stop and speak; don't speak on the move. Use movement between stopping points to answer questions, but do not irritate people by imparting important information out of their hearing. Stop and wait for your party to catch up. Point out features of interest before you get to them, not afterwards.

> **Inaugural lectures: Put your work into its institutional and intellectual framework.**

In academic institutions, it is common for each new professor to give an inaugural lecture. Such lectures are rites of passage (in the sense known to anthropologists) celebrating the transition of an individual from one state of being to another. Like all such ceremonies, their function is not only to celebrate the new status of an individual but also to celebrate the group in which that status is enjoyed.

Usually, inaugural lectures are given in public to audiences of widely varied technical backgrounds. The suggestion given in Chapter 3 for dealing with such situations are particularly relevant here. But whatever strategy you adopt when you give your inaugural lecture (or, a very similar situation, receive your medal), you may wish to do some of those things usual in rites of passage (cf. weddings and funerals), namely celebrate the past (by reflecting on some of the key individuals who shaped your discipline); recognise the present (by giving credit to colleagues and support staff); look to the future (by describing some of the directions in which you propose to move). Combined with a thesis (stated at the start of the lecture) about what might constitute the most important development to be taken over the next "x" years, such a celebratory approach is highly appropriate for a mixed audience.

> **Industrial presentations: Emphasise differences.**

It has become fashionable to accompany some types of industrial presentation with clouds of dry-ice "smoke", criss-crossed with flickering laser beams, to the accompaniment of symphony orchestras. Such goings-on seem to attract TV coverage, (though I, for one, groan when yet another flight of balloons straggles into the sky). I am not here dealing with that type of presentation, but rather with presentations in which technical specialists explain innovative ideas to those they hope will adopt them — either as financial backers or as purchasers.

Industrial presentations have been at the heart of the issues discussed in this booklet; little extra space will, therefore, be devoted to them here. Two points do, however, merit emphasis.

First industrial presentations often need to concentrate on explanations so that people can feel confident that they are making informed choices.

Second, whichever strategy from Chapter 3 you use, you may wish to concentrate on differences — those factors which distinguish your system, device, procedure from previous ones.

Preliminary paperwork can be used to highlight differences by listing factors in parallel so that points of similarity with previous systems, devices, and procedures are made plain. Careful use of different typefaces and/or colour can then highlight the differences which can be further stressed in your presentation.

Industrial tourism: Keep it simple.

In recent years, many industries have begun to discover the value of opening their doors to the public. Firms which make relatively small items of high value (such as bottles of drink or china) find attractive opportunities for direct selling; other industries seek to establish a brand image or generally improve public relations.

Many firms have found that an industrial tourism facility, far from being a drain on resources, can be a self-financing or even profit-making concern. Even without these attractions, as the demographic decline starts to bite, firms will no doubt find it increasingly attractive to show young people around with a view to attracting them into employment.

Abundant advice is now available (for example from the English Tourist Board) on the economics of adapting a work site as a tourist resource. There are many issues to be considered by any firm contemplating starting in on industrial tourism. For example, can you afford it? Can you afford not to do it? Should visitors wander freely or should they be guided? If wandering freely, should they be supplied with radio wands or cassette tape-recorders, or left with display boards, tape-slide machines, closed-loop film-projectors, or video? If guided, should they be guided by specialist guides? by non-specialist guides? by students?

As someone with a weakness for visiting industrial sites, I have suffered from presentations which left me dizzy from a stream of undifferentiated and unsorted information. The organisations which do the job well follow one key maxim: keep it simple. Rather than trying to say everything, good practitioners tell a story — usually emphasising one idea, perhaps the procedure that imparts unique value to the product.

To guide unaccompanied visitors, several levels of interest can be addressed by well-designed display boards — heading and photo for basic idea; a few lines in large bold type for school parties; a more detailed explanation in smaller type for specialist visitors. Guide books can reproduce this material for longer-term record.

For visits involving presentations, the rules are basically similar to those for coping with occasional visitors. The principal difference is that the planners will probably not be able to find out much more about the visitors except in very general terms. For this reason, a strategy is required which has meat in it for specialists — but serves as a basic guide to the non-specialist.

The trick is to pick some feature of the plant or process for special attention, to emphasise its significance during your introduction in the holding area beforehand, and then to use it as a focus for attention during the presentation. With a little care, visitors will go away with one central idea around which to arrange all the others.

> **Lectures to students: Concentrate on providing incentives for, and advice about, learning.**

The objective of teaching is to facilitate learning.

This may seem shatteringly obvious, but some teaching in higher education drifts into a form of presentation in which the purpose of the excercise becomes lost to sight. The phrase "covering the ground" can be a symptom of this — sometimes revealing that the lectures are planned in terms of what the lecturer wishes to say rather than in terms of what the students need to learn.

I have already advanced the notion that university education consists of materials, advice, and assessment. Problems arise because lectures often constitute a mixture of materials and advice. There may be merit in thinking of these elements separately.

A common pattern of advanced technical education consists of lectures supported by group tutorials. The tutorials are designed to offer advice to help students to grasp difficult lecture material. Apart from being costly in staff time, tutorials are often poorly attended — probably because the advice is distant in time from when it is needed. Might it not be better to think of the combination of lectures and tutorials as a single operation?

Classes based on guided independent study (PSI or Keller-plan materials), have been shown to be not only effective but also to be highly popular with students. Lecturing does indeed take place in such circumstances, but rather than being the primary method through which content is communicated, it becomes much more like series of self-contained, one-shot presentations of which the fundamental purpose is to excite interest and attention and to give students an overview of terrain rather than a guided tour of every hill and valley, or to "trouble-shoot" on concepts which groups of students have found difficult.

In a system of this sort, the plan is embodied in the paperwork — which may in turn be keyed into books and papers. You must, nevertheless, follow the axioms proposed in Chapters 2 and 3 to explain the purpose and plan of each presentation and also to show where it fits into the overall plan of the course of instruction. With a little ingenuity, it should be possible to use each of the strategies proposed in Chapter 3 and thereby show how specific subject-matter can be thought of in a variety of contexts (research, development, selling, maintenance).

It is also possible to get feedback as you go along, even from very large classes (see the section on managing questions in large meetings in Chapter 9 below).

> **Presentations to funding bodies: Remember that your judges are generalists.**

Most of the points concerning industrial presentations apply with equal force to presentations to funding bodies. There is, however, one point that requires special emphasis, namely that those who make decisions about research grant applications, although distinguished specialists in their own fields, are operating as generalists in making their judgements. Although they may possess considerable knowledge about the sub-field in which your topic is located, they are there to make judgements between different sub-fields (or even fields) and cannot possibly know everything.

Even specialists need an overview to point up the significance of what is proposed. Both your written submission and your presentation need to concentrate on highlighting rather than on minutiae of procedure. The details of procedure may well be asked for in the body of the research proposal, but a spoken presentation is more like an abstract — focusing on key points.

> **Project reports: Think of your listeners asking "So what?"**

Many university departments of science and engineering now ask students to give spoken presentations, as well as written reports, on their projects. Such presentations can be immensely rewarding to all concerned — but only if properly thought through.

What action will be taken by whom on your project report? Answer: My professor will mark it. Gloom for you; gloom for the professor. But the situation can be transformed if you think of your professor as an editor who is helping you to polish your report for publication. Both the professor and you yourself will enjoy the work if it builds towards an industrial presentation. See if you can arrange to send the report to someone who will take some action on it. Invite that person to the spoken presentation so that you are genuinely speaking to some purpose — namely that of changing someone's behaviour.

The modest expedient of making college presentations "for real" will help you to decide the content: what are the key findings? what is their theoretical and practical significance? who is likely to be affected by what you have found out/made? what should they do?

> ## Public inquiries: Be prepared to improvise.

Public inquiries are often pretty noisy affairs with feelings running high, and strong positions being taken by participants. They are political occasions where people may be more interested in saying what they think than in listening to what you have to say.

If you have the opportunity to give a short opening presentation, the suggestions offered earlier in this booklet should give you ideas on what to do. You may, however, then have to give a series of mini-presentations on issues raised by participants, i.e. more than just brief answers to questions. You will obviously be able to do this well if you have researched the issues which are bothering people.

Often, these will centre on perceptions of risk or loss of amenity. Identify areas of dispute by reading local papers, listening to local radio, talking in pubs, and where necessary, commissioning or carrying out surveys of views. Your presentation can then be driven by the questions on people's minds.

Be prepared to improvise: this is more impressive than simply delivering a prepared statement. If you can gather questions (see Chapter 9 below) and respond to them systematically and thoroughly, you will be addressing matters which *actually* concern people rather than those that you think may concern them.

At the practical level, whether or not you have used slides, film, and video in your opening presentation, in the cut-and-thrust of a question-driven session you will almost certainly find an overhead projector or Multi-Media Projector more convenient than a 35mm slide projector. Take plenty of (simple) prepared transparencies — and plenty of blank transparencies and pens.

School talks: Don't talk down to children.

With increasing fears about under-recruitment into professions and into higher education institutions, growing numbers of professionals are giving talks in schools. In addition to the suggestions already offered in this booklet, one further principle should be asserted: do not talk down to children.

Rather than generalised exhortation, they want detail — some insight into different types of work. Therefore, don't sell institutions: sell ideas. Children are shrewd: if they are to become interested in higher education, they will be looking for intellectual stimulus. So, talk about some of the most challenging aspects of the work. Better still, describe a gritty problem, and ask the pupils how they would tackle it. A few minutes spent by children talking together in groups to come up with an answer to the problem you have posed will be more effective than the same amount of time listening to you — however engaging you are! They are not going to a funfair; so, rather than talk about the recreation facilities, speak technically.

Section meetings: Link the new to the old.

A very common form of technical presentation is that involving people from different sections or divisions in an organisation, so that everyone can keep up to date with what is going on. Meetings within sections are also held at which individuals or small teams describe their work.

When people meet regularly, it may seem self-evident why they are there, but never assume this. Before launching into details about some new project or programme, recap where you are. Link the agenda with that of previous meetings to establish a framework of shared expectation. Quite apart from coping effectively with any change of staff, or lack of information from absences, a range-finding introduction (after an initial statement of your purpose and plan), can ensure that colleagues see how your interests key in with theirs. A few minutes spent "locating" your work within the broad context of your organisation's activity can not only give busy people time to "tune in to your wavelength", but can also sharpen interest considerably. Indeed, sometimes the discipline of thinking these relationships through can be an important hidden agenda of meetings.

Telephone calls and consultations.

Plan your material as though for a presentation. It may seem strange to find observations about talking to individuals in a booklet concerned with public presentations. However, some encounters which may seem like conversations are sufficiently formal (and costly) to merit special treatment.

Sometimes when you receive a telephone call you are in effect receiving an invitation to give an impromptu technical presentation. Likewise, when you call a colleague, you may well be giving a mini-presentation (nowadays probably with the benefit of facsimile-transmitted diagrams or interactive TV). In both receiving and and making calls, it is essential to plan what you have to say.

When you ring a colleague, identify yourself, make a greeting, and then — just as in a public presentation — start straight in with a statement of your purpose in telephoning. Then rapidly establish any necessary framework of reference (for example by sketching in who, what, when, why, where, how and with what effect) before getting to the detail on which you need specific action. Again, think in terms of what you want the person to do.

When you receive a call, you may need to listen carefully to discern what your caller wants. Try to get inside the caller's position, and then structure your reply in an orderly fashion.

Consultations likewise require careful presentation. If you are seeing people rapidly in succession, read notes beforehand — even if you have to keep your client/patient waiting for an extra minute. It is very off-putting to be confronted by a professional fumbling through notes muttering "You are 'er ..." and clearly having no clear perception of why you are there!

Many consultations can be driven by questions, but people may not know what they want to ask. That is why they have come to see you. Draw on experience to phrase their thoughts and feelings: "You may be wondering/feeling…" Anticipate matters, such as cost, hassle, safety, likely to be on their minds even if they do not mention them. Do not assume knowledge that they may not possess. To avoid patronising, use phrases like: "As you know…" If they do not know, you will go on to give the necessary information; if they do know, they will feel pleased that you recognise this.

Chapter 9

Chairing Sessions and Managing Questions in Large Meetings

Many technical presentations are chaired or have question and answer sessions or both. This chapter is written on the assumption that you are chairing the meeting. If you are speaking, you should be able to expect these services from the person chairing your meeting. If no-one is in the chair, you will need to attend to these matters yourself.

> When you chair a technical presentation, you assume all the responsibilities of a host. You need to know not only your principal guest (the speaker) but all the services that both speaker and audience (your other guests) may need.

Before the meeting

- Get and read the full texts of the papers to be presented, to plan effective introductions, continuity and review.

- Identify in advance likely questions and areas of agreement/ disagreement, and plan questions/discussion sessions with the speaker(s).

- Find out about necessary facilities and services: emergency and technical services; first aid; light switches; fire exits; alarm procedures; loos; times and places of catering.

- With the speaker(s), check that any necessary apparatus (microphones, slide-projectors, air-conditioning, etc.) are working properly and if they are not, call help.

At the meeting

- Introduce the speaker(s) by name and say a few words about her/him/them. Use one or two anecdotes to individualise; but be brief (not full biography or flattery). Avoid clichés. Pronounce a speaker's name the way the speaker likes to have it spoken.

- State the topic and say why it is important/interesting/timely.

- Help the speaker(s) to keep to the allotted time; be strict about this!

- End the session on time by thanking the speaker(s).

Managing questions in large meetings

- Have questions to ask/points to make yourself, in case the follow-up to the opening presentation is slow to warm up.

- Ask people to give their names and to be brief in their remarks so that as many people as possible may contribute.

- Ask for a show of hands to see how many people want to ask questions/make points.

- Take a bunch of questions/points at a time (rather than have the speaker(s) respond to each) and ask people with relevant supplementary points to show hands while issues are being discussed.

- Write questions up on the OHP and get a quick show of hands to find an order of priority for having them answered.

- Try to take questions/points from different people rather than letting one or two people dominate the discussion — unless some issue of obviously great importance is being pursued.

- Towards the end of the session, steer the discussion round to the question of what action should be taken by whom (if this has not happened spontaneously).

- Stop while questions are still coming, rather than let a meeting fade away.

If you are the person being questioned...

- Do not accept questions during your presentation: this will wreck your plan. Say politely but firmly that you will be delighted to answer questions afterwards.

- Be brief in your answers. People will ask for more information if they want it.

- Do not give serious answers to joke questions: smile and invite the next question.

- If you do not know the answer to a question, say so: never try to bluff your way through. Offer to send information if the questioner will give you an e-mail address after the meeting, or ask if anyone else present can help. Stern counsel: but no-one expects you to know everything — and your reputation will be enhanced if you make it clear that you are more interested in your listeners getting information than in "saving face".

- Never "put people down". Invite the "all-knowing questioner" to speak. Use an obvious expert positively: invite others to ask questions of him/her.

- If you are confronted by a person who talks too much, write that person's questions on the OHP; invite others to join in.

- If someone tries to dominate proceedings, be firm but always polite and try to draw others in. With the person who just takes time to formulate a question, intervene politely and try to boil the question down.

- Where a questioner is clearly antagonistic, restate the issue so that he/she sees you have taken their point of view on board.

> **Build question-and-answer sessions into presentations to encourage feedback.**

In one-shot presentations, you may wish to have a brief introductory session followed by a longer session driven by the questions of those present. Rather than ask: "Are there any questions?" and have an awkward silence, make clear that there will be a question period; immediately after your opening remarks, invite people to talk with their neighbours to decide on important questions (turn your back — and gather notes, OHP transparencies, etc. — to signify that you are leaving time for this process of mutual consultation to take place); gather questions on the OHP as indicated above; answer first the questions of interest to the greatest number of people.

In lectures in a university course, the use of this procedure (sometimes known as the "buzz group technique") can help students to pay attention to your discourse by providing periods of varied activity; it can also help you by providing periodic checks on how well you are being understood.

A good question-and-answer session is the surest sign of an effective presentation because

> **In answering questions, you are forced to address what your listeners want to know rather than what you want to say.**

When you answer questions in a conversation, you speak naturally in response to what your listener wants to know. No doubt you find yourself saying: "Speaking technically, I would say…"

Appendix A

Count-Down Check List

This list contains contains a number of points which you will wish to check when you are first invited to give a presentation, and in the planning stage, and immediately before you give it.

> **When you are invited to speak, find out as soon as possible:**

- The exact title by which your presentation will be advertised;

- Why you were chosen and what sort of approach is expected of you;

- The nature of the occasion;

- Who will be there (age, status, numbers);

- Why they will be there;

- Their attitude to your subject (if your audience will be, for example, a hostile pressure group, you need to know this);

- How you should travel to the site;

- Who will pay your fee (if any) and/or expenses: whether there is a form;

- If an overnight stay is involved, who will make the arrangements and where will you be in relation to where the presentations is to be given;

- Name and telephone number of someone to contact on the day in case of travel trouble; illness or other difficulties;

- Whether there will be food before or after the presentation (if you have special requirements, make them known);

- When you can see the room in which your presentation is to be given;

- What Audio-Visual facilities will be available:
  - Technical details (such as screen settings) of Multi-Media Projector;
  - Type of projector — rack or carousel;
  - Projectionist or not;
  - Microphone or not: If yes, fixed or neck.

- For Multi-Media Projection, see if you can send your computer file in advance by e-mail — so that it can be loaded into the local computer and tried out before you arrive.

- If you like a carafe of water, ask for it;

- If your presentation is one in a series, what previous presentations have covered;

- For how long you will be expected to speak;

- What is the start time;

- Whether there will be questions and discussion;

- Whether there will be a lectern/table/desk (assume, however, that there will be none and take a clip-board);

- What preliminary paperwork will be needed and by when (abstract, a full paper, discussion notes).

In the planning stage…

- Organise the production of slides, video, film, flip-chart, graphics, etc. as soon as possible to avoid last-minute crises.

When you get to the site…

- Walk round the room;

- Try out your slides/overhead projector transparencies/video/ film and try to view from different seats (if the slides cannot be seen properly, take whatever remedial action is possible);

- If you have a walk to a rostrum, try it out to note cables, steps, or other hazards that might foul things up for you;

- Check the locations of: pointer; water; switches (lights, microphone, slide projector, OHP);

- Meet the projectionist and discuss signals to be given or how to work the auto-changer;

- Check the ventilation to ensure that the room is not too hot and stuffy or too cold, and find how to alter matters during your presentation if the need arises.

Write at the top of your notes items to check before you start speaking.

Before you are introduced (or, if you are alone, before you signal the formal start of your presentation), make sure that everything is as you would want it. Write a mini check list at the top of your notes. For example:

- Where is the pointer?
- Where is the slide changer?
- Is the microphone on or off?
- Is the height and position of the microphone satisfactory?
- Is the OHP still in focus?

Appendix B

Notes to Accompany the Speaking Technically Presentations Check Sheet

These brief notes pick up some of the points raised in the booklet and relate them to the headings of the Presentations Check Sheet (Appendix C).

A. CONTENT AND ORGANISATION

1. Awareness of audience knowledge and interests

- Did the speaker say how the presentation related to your interests and needs?

- Did the speaker explain the purpose of the presentation in terms of action that you might take?

2. Pacing of material

- Was the amount of information appropriate for the time?

- Were measures taken (such as section summaries) to ensure that you had understood important ideas as the presentation proceeded?

3. Timing

- Delete one mark on the check sheet for every minute over or under the stated time of the talk.

4. Clarity and organisation

- Was the main point of the presentation clear?

- Was the material logically and coherently organised? If not, how could it have been improved?

5. Theme/Dramatic shape

- Did the speaker explain how the talk was organised (in addition to stating its purpose)?

- Did the speaker use an effective strategy in organising the material? If not, what would have been a better strategy?

- Did the speaker build around a series of questions rather than simply make statements?

6. Emphasis of important points

- Did the speaker highlight key points with visual aids or by saying that they were important?

- Did the speaker indicate clearly what was not important, for example by indicating clearly what was background material?

7. Summary

- Did the speaker leave you with a clear understanding of the main point of the talk?

B. THE SPEAKER

8. Audibility

- If the speaker was inaudible, why was this?

9. Quality

- Did the speaker vary the tone and volume of his/her voice, for example by slowing down to emphasise particularly important points?

10. Speed of speaking

(To be distinguished from pacing of material which is dealt with in section 2) above:

- At which points (if any) did the speaker speak too slowly or too quickly?

11. General manner

- If the speaker's general manner was in any way unsatisfactory, how could it have been improved?

- If the speaker's manner was particularly agreeable, why was this?

12. Posture

- If the speaker was awkward and tense, or sloppy and chaotic, what should be done to improve matters?

13. Movement

- If the speaker moved about too much or too little, what specific remedies could be applied?

14. Gestures

- Did the speaker use his/her hands effectively? If not, when and where could specific gestures have been included?

15. Eye contact

- Did the speaker read from a script, seem too note-bound, or lose track in his/her notes?
- Was eye-contact appropriate?

C. USE OF VISUAL AIDS

16. Positive use of visual aids

Did the speaker's visual aids:

- Help to emphasise important points?
- Help to pace the material?
- Appear legible/clearly visible?
- Have the right amount of detail?
- Add interest to the presentation?

- Stay on view long enough?
- Other (specify)

17. Damaging use (or lack) of visual aids

Did the speaker's visual aids:

- Distract attention from the presentation?
- Slow the presentation down too much?
- Appear illegible/hard to see?
- Have too much detail?
- Add nothing to the presentation/seem boring?
- Stay on view for an inadequate time?
- Did the presentation need additional visual aids? (specify)

Appendix C

Speaking Technically
Presentations Check Sheet

For each section, please circle the statement which you think applies and write the numerical score in the space at the right of the sheet

A. Content and Organisation

1. *Awareness of audience knowledge and interests*

Clear statement of the purpose of the talk and how it related to your interests.

Weak introduction.

No statement of the purpose of the talk or how it related to your interest

10	9	8	7	6	5	4	3	2	1	0	Score

2. *Pacing of material*

| Ideas and facts too thinly spread. Boring. | Ideas and facts well paced: easy to follow. | Ideas and facts too densely packed. Confusing. | ☐ |

0 1 2 3 4 5 6 7 8 9 10 9 8 7 6 5 4 3 2 1 0   Score

3. *Amount of material*

| Too little (under-ran time). | Satisfactory. | Too much (over-ran time) | ☐ |

0   1   2   3   4   5   4   3   2   1   0   Score

4. *Clarity and organisation*

| Material logically organised and coherent. | Material weakly organised. | Material badly organised. | ☐ |

10   9   8   7   6   5   4   3   2   1   0   Score

5. *Theme/Dramatic shape*

| Speaker explained how talk was organised and/or gave it some dramatic shape & theme. | | Speaker did not explain how talk was organised. Shapeless talk. | ☐ |

5   4   3   2   1   0   Score

6. *Emphasis*

Speaker emphasised important points as he went along.		Inadequate emphasis.		Speaker did not emphasise important points.		

5	4	3	2	1	0	Score

7. *Summary*

Excellent summary of main points.		Inadequate summary.		No summary of main points.		

5	4	3	2	1	0	Score

MAXIMUM POSSIBLE MARKS ON CONTENT AND ORGANISATION: 50   **TOTAL SCORE**

B. The Speaker

8. *Audibility*

Clear.		Difficult to hear.		Inaudible.		
5	4	3	2	1	0	Score

9. *Quality*

Lively/Varied Tone.		Satisfactory.		Dull.		
5	4	3	2	1	0	Score

10. *Speed*

Too fast.		About right.		Too slow.		
0	1	2		1	0	Score

11. *General Manner*

Agreeable.		Satisfactory.		Poor.		
5	4	3	2	1	0	Score

12. *Posture*

Awkward and tense.			Relaxed.		Sloppy.		
0	1	2	3	2	1	0	Score

13. *Movement*

Too static.   OK.   Too much.

| 0 | 1 | 2 | 1 | 0 | Score |

14. *Gestures*

Effective/natural.   Satisfactory.   Awkward.

| 3 | 2 | 1 | 0 | Score |

15. *Eye Contact*

Insufficient.   Satisfactory.   Too much.

| 0 | 1 | 2 | 3 | 4 | 5 | 4 | 3 | 2 | 1 | 0 | Score |

Write Minus scores if speaker :
Read talk from script
(Minus 5)

Was too note-bound
(Minus 4)

Lost track in notes   Score
(Minus 3)

Maximum possible marks on speaker : 30

**Total
Score**

C. Use of Visual Aids

16. *Positive use of visual aids*
   (Blackboard, Slides, Film, Exhibits, Demonstrations etc.)

Add marks — up to 4 marks per item, maximum +20:

> • Helped to emphasise important points.
> • Helped to pace material.
> • Legible/Clearly visible.
> • Added interest to talk.
> • Adequate time allowed for us to see.
> • Other (Specify)

$+$ ☐

Score

17. *Damaging use of visual aids*

Subtract marks — up to 4 marks per item, maximum −20

> • Distracted attention from the talk.
> • Illegible/hard to see.
> • Too much detail.
> • Added nothing to the talk/Boring.
> • Inadequate time for us to see.
> • Talk badly needed visual aids.
>   (Specify which sort)

$-$ ☐

Score

Maximum possible marks on Visual Aids : 20   **Total Score**   ☐

Comments   ☐

Adjustment for overall impression, +10 or −10   Score

Grand Total Score (Maximum 100)   ☐